GREAT WHITE SHARK

By Rachel Rose

Consultant: Erin McCombs
Educator, Aquarium of the Pacific

BEARPORT
PUBLISHING

Minneapolis, Minnesota

Credits

Cover and title page, © Stephen Frink/Getty, © Kvitka Nastroyu/Shutterstock, © Yuri Samsonov/Shutterstock; 3, © nabil refaat/Shutterstock; 4-5, © Nature Picture Library/Alamy; 6-7, © Johan Holmdahl/Shutterstock; 8-9, © RamonCarretero/iStockPhoto; 10-11, © SylwiaDomaradzka/iStockPhoto; 12-13, © Wildestanimal/Alamy; 14, © GlobalP/iStockPhoto; 14-15, © AlbertoLoyo/iStockPhoto; 16, © abadonian/iStockPhoto; 16-17, © Luke Sorensen /Alamy; 18-19, © EMPPhotography/iStockPhoto; 20-21, © cdascher/iStockPhoto; 22, © Ramon Carretero/Shutterstock; 22-23, © nabil refaat/Shutterstock; 24, © nabil refaat/Shutterstock.

President: Jen Jenson
Director of Product Development: Spencer Brinker
Senior Editor: Allison Juda
Associate Editor: Charly Haley
Designer: Colin O'Dea

Library of Congress Cataloging-in-Publication Data

Names: Rose, Rachel, 1968- author.
Title: Great white shark / by Rachel Rose.
Description: Minneapolis, Minnesota : Bearport Publishing Company, [2022] |
    Series: Shark shock! | Includes bibliographical references and index.
Identifiers: LCCN 2021026718 (print) | LCCN 2021026719 (ebook) | ISBN
    9781636915319 (library binding) | ISBN 9781636915401 (paperback) | ISBN
    9781636915494 (ebook)
Subjects: LCSH: White shark--Juvenile literature.
Classification: LCC QL638.95.L3 R65 2022 (print) | LCC QL638.95.L3
    (ebook) | DDC 597.3/3--dc23
LC record available at https://lccn.loc.gov/2021026718
LC ebook record available at https://lccn.loc.gov/2021026719

For more information, write to Bearport Publishing, 5357 Penn Avenue South, Minneapolis, MN 55419.

CONTENTS

Gotcha! 4

Keep On Swimmin' 6

Speedy Giants. 8

A Great Bite 10

On the Hunt 12

Fat Food 14

Danger for Great Whites. 16

Great White Pups 18

Long Life 20

More about Great White Sharks 22

Glossary 23

Index. 24

Read More. 24

Learn More Online 24

About the Author. 24

GOTCHA!

A great white shark hunts for its next meal in ocean waters off the **coast**. Suddenly, it spots a seal near the **surface** of the water. The huge shark slowly moves underneath the seal, then rushes up in a sudden burst. The shark catches the seal in its giant jaws.

Great white sharks don't chew their food. They rip their **prey** in large chunks and swallow the pieces.

KEEP ON SWIMMIN'

Great white sharks, also known as white sharks, live in most of the oceans around the world. They can be found along the coasts of North America, South America, Australia, Asia, and Africa. These sharks are always on the move, usually to search for food.

White sharks **migrate** for long distances. Some have been tracked swimming from South Africa all the way to Australia.

GREAT WHITE SHARKS AROUND THE WORLD

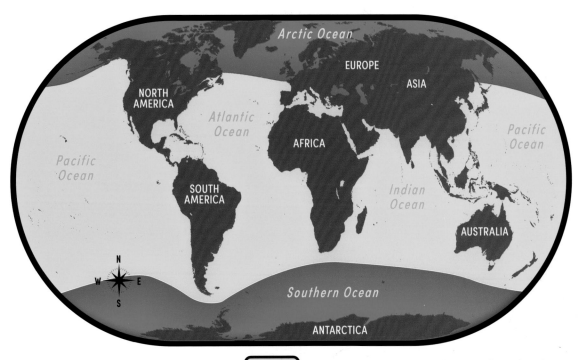

Where great white sharks live

Scientists once tracked a white shark they named Lydia. The shark swam more than 45,000 miles (72,420 km) in 5 years.

SPEEDY GIANTS

These sharks make their way through the ocean quickly. Great whites can swim up to 35 miles per hour (56 kph).

And the fast swimmers are among the biggest fish in the ocean. The largest great whites weigh up to 5,000 pounds (2,270 kg). That's as heavy as an SUV!

Great white sharks can be more than 20 feet (6 m) long.

A GREAT BITE

Look at that mouth! These big sharks have about 300 extremely sharp teeth. The teeth are arranged in as many as seven jagged rows. The front rows are used for biting and ripping into prey. When those front teeth fall out, the rows farther back move forward to replace them.

Great whites have powerful jaws. Unlike humans, both their upper and lower jaws can move.

On the Hunt

It's not hard for white sharks to find something to sink their many teeth into. Great whites are excellent hunters. Their strong sense of smell helps them easily find prey. They can smell a drop of blood in the water from more than 1,500 ft (460 m) away.

These sharks have great sight, too. They can see their prey day or night.

Great whites can roll their eyes back in their heads. This **protects** their eyes from being scratched by prey.

FAT FOOD

What exactly are great white sharks hunting for? Sometimes, they eat other fish. But mostly, white sharks eat seals and sea lions. These animals have a lot of fat called blubber on their bodies. After a great white shark eats a good blubbery meal, it doesn't need to eat again for days.

Great whites have also been known to eat whales, dolphins, rays, and sea turtles.

DANGER FOR GREAT WHITES

Great white sharks are among the strongest creatures in the ocean. But that does not mean they're always safe. These sharks are often caught and killed in large fishing nets by accident. There are also some people who hunt these animals, even though many countries have **banned** fishing for great white sharks.

People can try to help white sharks by getting seafood only from companies that are careful about how they fish.

GREAT WHITE PUPS

Great whites are being hunted and killed at a faster rate than they are having young. This means the number of white sharks in the world is getting smaller.

It takes about a full year for baby great whites to grow inside their mothers. The mother sharks may have 2 to 10 babies at a time. These baby sharks are called **pups**.

Baby white sharks are usually about 5 ft (1.5 m) long when they are born.

Long Life

As soon as they are born, great white pups have to look after themselves. Sometimes, they are eaten by big fish or other sharks. The pups that **survive** eat other fish until they grow big enough to eat larger prey, such as seals. White sharks grow for several years before they reach their full size.

> Great white sharks can live for up to 70 years.

More about
Great White Sharks

Great whites can rip off 30 lbs (14 kg) of meat from their prey in one bite.

Like all sharks, great whites don't have bones. They have flexible **cartilage**, which helps them swim with smooth movements.

White sharks have strong, powerful tails that help them swim fast.

Great whites are sometimes hunted by killer whales.

White sharks are named for their white bellies.

Like other fish, white sharks breathe by having water move through their **gills**.

GLOSSARY

banned not allowed

cartilage strong, rubbery material that makes up a shark's skeleton instead of bones

coast an area where land meets an ocean

gills the parts of a fish that allow it to breathe underwater

migrate to move from one place to another at a certain time of the year

prey animals that are hunted and eaten by other animals

protects keeps safe

pups baby sharks

surface the top or the outside of something, such as the top of ocean water

survive to stay alive

Index

eyes 13
fins 16–17
humans 10, 15–16
hunt 4, 12, 14, 16, 18, 22
jaws 4, 10
oceans 4, 6–7, 9, 16

prey 4, 10, 12–13, 20, 22
pups 18–20
seals 4, 14–15, 20
sea lions 14
smell 12
teeth 10, 12

Read More

Gray, Susan H. *Great White Shark Migration (Marvelous Migrations).* Ann Arbor, MI: Cherry Lake Publishing, 2021.

Nixon, Madeline. *Great White Shark (Sharks).* New York: AV2, 2019.

Learn More Online

1. Go to **www.factsurfer.com** or scan the QR code below.
2. Enter "**Great White Shark**" into the search box.
3. Click on the cover of this book to see a list of websites.

About the Author

Rachel Rose lives in California. She swims in the ocean every day, and she sees plenty of seals—but she hasn't seen a shark yet!